Landforms

CAVES

by Sonja Olson

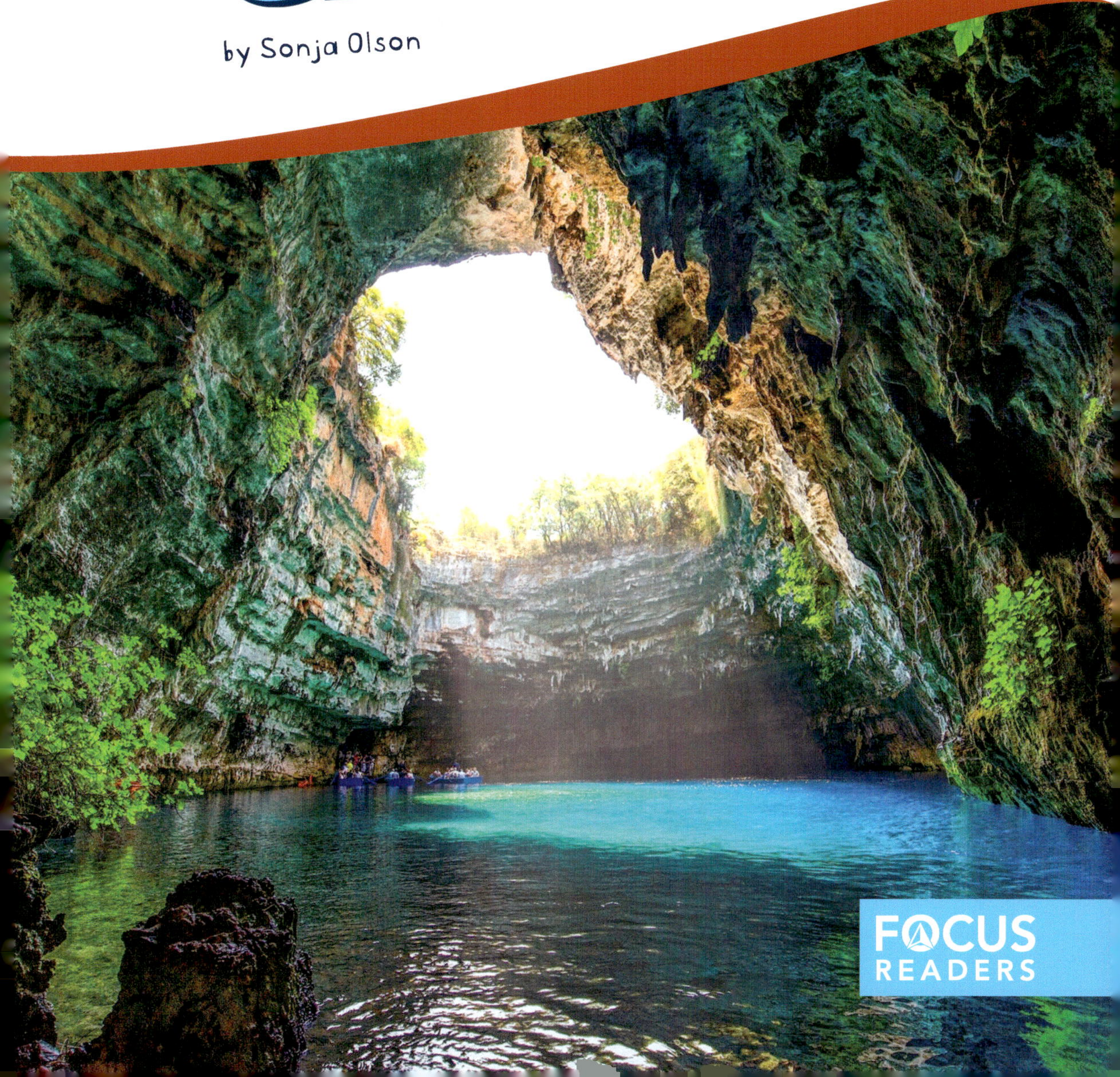

FOCUS READERS

FOCUS READERS

www.focusreaders.com

Focus Readers is distributed by North Star Editions:
sales@northstareditions.com | 888-417-0195

Produced for Focus Readers by Red Line Editorial.

Photographs ©: Piotr Krzeslak/Shutterstock Images, cover, 1; JianweiZ/iStockphoto, 4–5; Stephen Alvarez/National Geographic/Getty Images, 7; jiawangkun/Shutterstock Images, 8–9; Nickolay Stanev/Shutterstock Images, 10; J. Heglason/Shutterstock Images, 13; siete_vidas/Shutterstock Images, 14–15, 29; diegograndi/iStockphoto, 16; TayHam Photography/Shutterstock Images, 18; Vietnam Stock Images/Shutterstock Images, 20–21; Dsdesign/Shutterstock Images, 22–23; Hendra Xu/Shutterstock Images, 24; gary yim/Shutterstock Images, 27

ISBN
978-1-63517-891-3 (hardcover)
978-1-63517-992-7 (paperback)
978-1-64185-195-4 (ebook pdf)
978-1-64185-094-0 (hosted ebook)

Library of Congress Control Number: 2018931688

Printed in the United States of America
Mankato, MN
May, 2018

About the Author

Sonja Olson obtained her bachelor of fine arts in sculpture from the College of Visual Arts and a master's in curriculum and instruction from the University of Saint Thomas. Prone to rescue stray cats, Olson currently teaches, writes, creates, and resides in Saint Paul, Minnesota.

TABLE OF CONTENTS

CHAPTER 1

UNDERGROUND WONDERS

Hundreds of **stalactites** hang from the ceiling. They look like icicles. But they are solid rock. Water drips from the stalactites. It falls into a shallow pool. Each drop makes a spooky echo in the cave.

Crystal Cave in Bermuda has huge stalactites that reach toward blue pools.

A cave is an empty space in the ground. Many caves are dark and cold. They can also be long and deep. Most caves take thousands of years to form. And once they are formed, most caves are always changing.

FUN FACT

The world's deepest known cave is the Krubera Cave. It is in the country of Georgia. It is more than 7,200 feet (2,195 m) deep.

A caver enters the Krubera Cave.

CHAPTER 2

WEARING DOWN THE EARTH

There are several types of caves. Solution caves are common. When this kind forms, **groundwater** flows through cracks in rock. The water slowly dissolves the rock. Over time, an empty space forms.

The Mammoth Cave System is a group of solution caves in Kentucky.

The Thurston Lava Tube is a lava cave in Hawaii.

Lava caves form after a volcano erupts. Lava burns a path into the ground. The lava on top cools and hardens. But the lava below stays hot. This hot lava keeps moving down the side of the volcano. It leaves behind a hollow space under the hard lava. This process forms a tunnel-like cave.

Waves can make sea caves. They form as waves crash against a cliff.

FUN FACT

Washington has many lava caves. Ape Cave is in Washington. It is approximately 2.4 miles (3.9 km) long.

This happens over and over. The waves **erode** the cliff over time. After many years, a cave is formed.

In cold places, caves can form in thick ice. When seasons change, the weather becomes warmer. Some of the ice melts. A hole forms in the melting ice. As more ice melts, the

FUN FACT

The world's longest cave system is in Kentucky. The Mammoth Cave System stretches more than 345 miles (555 km).

This glacier cave is in Iceland.

hole gets bigger. Over time, a cave forms. This is called a glacier cave.

CHAPTER 3

ALWAYS CHANGING

Caves change over time. They become larger as more rock erodes. Many caves stay damp all the time. In these caves, **minerals** from the water stick to the rock. They keep the thin ceiling strong.

Sea caves grow larger as waves crash against the rock.

The hole caused by a collapsed cave ceiling is called a cenote.

If a cave gets too dry, the ceiling can fall down. This creates a big hole that is open to the outside air.

More light gets in. This changes what plants are able to grow in the cave. It also changes what animals can live in the cave.

Many caves are made out of rock. They may seem very strong. But they are **fragile**. Humans sometimes cause harmful changes in these caves. For example, touching a stalactite can affect how it grows. Humans have germs and oil on their hands. These things can damage the rock.

Humans can help protect caves by building walkways for visitors to use.

The germs can also make animals sick. **Pollution** and **climate change** put cave creatures in danger, too. The plants and animals that live in caves need the temperature to stay the same. They can be harmed if it changes. That's why it is important for people to protect caves.

FUN FACT

Some people explore caves for fun. This activity is called spelunking.

SON DOONG CAVE

Son Doong Cave is the largest known cave **passage** in the world. It is in the jungles of Vietnam. The cave was found in 1991.

Son Doong Cave is more than 5.5 miles (8.9 km) long. Some parts of the cave have large openings in the ceiling. These openings can reach heights of 400 feet (122 m).

In one area, the cave roof fell apart. This change allowed light to get in. Over time, this part of the cave became a jungle. Plants and trees grew underground. The trees grew to be nearly 100 feet (30 m) tall.

Plants grow in the Son Doong Cave now that light can get in.

CHAPTER 4

LIFE UNDERGROUND

Caves support many kinds of life. But each part of a cave is home to different plants and animals. There are three main parts of a cave.

The light zone is the beginning of the cave. It gets the most light.

Plants can grow where light reaches into the entrance of the cave.

Fruit bats hang from the ceiling of the Monfort Bat Cave in the Philippines.

Many ferns grow here. Raccoons and bears use it for shelter.

Next is the twilight zone. It is darker and cooler. Moss grows here. Salamanders eat insects that live on the cave walls. Bats also sleep in the twilight zone.

Bats use **echolocation**. It helps them fly through the dark cave. Bats make high-pitched sounds. These sounds bounce off the hard cave walls. The sounds come back to the bat's ears. The **reflected** sounds tell the bats how close they are to objects.

FUN FACT

Bats are important to other living things in caves. Their waste is food for spiders and insects. It also helps moss and fungi grow.

The last part of the cave is the dark zone. There is no light. No plants are able to grow. But spiders, shrimp, fish, snails, and worms all live here. These animals are blind. They use their hearing and feelers to move around.

FUN FACT

Bacteria found nowhere else on Earth can live in caves. Some look like jelly. Others look like fluffy white balls.

Glowworms live on the ceilings of the dark zone of the Waitomo Caves in New Zealand.

FOCUS ON

CAVES

Write your answers on a separate piece of paper.

1. Write a sentence that explains how a solution cave is formed.
2. What zone of a cave would you want to visit? Why?
3. What sense do animals use to move in the dark zone of a cave?
   - **A.** sight
   - **B.** touch
   - **C.** smell
4. How can people help protect caves?
   - **A.** Don't touch anything in the cave.
   - **B.** Take a cave animal as a pet.
   - **C.** Leave trash behind when exploring.

5. What does **dissolves** mean in this book?

*The water slowly **dissolves** the rock. Over time, an empty space forms.*

A. breaks down
B. turns to a liquid
C. flows through the cave

6. What does **erupts** mean in this book?

*Lava caves form after a volcano **erupts**. Lava burns a path into the ground.*

A. is not active
B. erodes away
C. becomes active

Answer key on page 32.

GLOSSARY

climate change
A long-term change in Earth's temperature and weather patterns.

echolocation
A way of bouncing sound off objects to figure out how close they are.

erode
To wear away slowly over time.

fragile
Easily cracked or broken.

groundwater
Water held underground in the soil or rock.

minerals
Substances that are naturally formed under the ground.

passage
An opening that people or animals can move through.

pollution
Harmful substances that collect in the air, water, or soil.

reflected
Bounced back from a hard surface.

stalactites
Icicle-like formations that hang from the ceiling of a cave.

TO LEARN MORE

BOOKS

Hamilton, S. L. *Caving*. Minneapolis: Abdo Publishing, 2014.

Labrecque, Ellen. *Caves*. Chicago: Capstone Heinemann Library, 2014.

Peter, Carsten, and Glen Phelan. *Extreme Planet: Carsten Peter's Adventures in Volcanoes, Caves, Canyons, Deserts, and Beyond!* Washington, DC: National Geographic, 2015.

NOTE TO EDUCATORS

Visit **www.focusreaders.com** to find lesson plans, activities, links, and other resources related to this title.

INDEX

Answer Key: 1. Answers will vary; **2.** Answers will vary; **3.** B; **4.** A; **5.** A; **6.** C